College

By

Marc Blackmon

Copyright 2015 by Marc Blackmon

All Rights Reserved

No part of this book may be reproduced in any form without written consent of the publisher.

ISBN 978-1506000497

Printed in the U.S.A.

Table of Contents

Chapter

1. Graduating High School — *Establishing an Identity*

2. Joining the Military — *Learning Who I Was*

3. The Ship Life — *Joining the Work Force*

4. Shore Duty Life — *Starting College*

5. Tough, Easy Life — *Learning to Be a Student*

6. Post-College — *Corporate America*

7. Paris, France — *A Different Perspective*

8. The Job Market — *A Down Economy*

9. Sydney, Australia — *Expatriate Life*

10. Postgraduate School — *Defining My Success*

Chapter 1- Establishing an Identity

It was May 19, 2000 and a bright, sunny day. This was a big day, and I had been looking forward to it for four years. I was finally graduating high school. High school seemed like it lasted forever. As a freshman, you tend to learn where you fit in. Class enrollment, clubs you join, and friends you make set the tone for the remainder of high school. I signed up for applied classes and did better than I expected. There were clubs to join but they did not really strike my interests.

Although I was a friendly guy, I was not a social butterfly. During my sophomore year, I ventured out and tried to make the junior varsity baseball team. It was a fun and memorable experience, but I did not make the cut. As a junior, my social life began to flourish as I studied for my driver's license, found a part-time job and began dating. This was the time when I began to think about my plans after high school. I did not care about attending college at the time because it seemed like a continuation of high school—but for another four years! I chose a different route and wanted to join the military. So I contacted a Navy recruiter and enrolled in the Delayed Entry Program. As a senior, I challenged myself academically, as I strived to leave my hometown and explore the world. This graduation day was well deserved and appreciated for all my studying, dedication, and hard work.

The year of 2000 was an exciting year. I had completed a life goal and was joining the United States Armed Forces. My last family gathering as a civilian was a barbeque. It was at a cousin's house with my great aunt, mother, brother, other family, and friends. I received so much praise for being ambitious and career-oriented. My recruiter came and picked me up late that night, and I was very anxious! I enjoyed meeting and eating with my relatives during this special time in my life. If there were a day I could go back and relive, it would be this day. Everyone was feeling patriotic, the food was fantastic, and there was plenty of love amongst us.

It was tough establishing an identity as a teenager because I had not been exposed to many adult situations. My parents made most of my decisions for me and dreaded me growing into adulthood. My parents were never married, and I lived with my mother during childhood and teenage years. My father picked me up as a child, but there was little interaction between us. I reached out to him when I turned sixteen years old for more guidance in life. My parents lived totally different lives. Both of them worked as blue-collar workers in factories and textiles, usually as machine operators. I did not want to be a machine operator and made a commitment to myself that I was going to move on and try a different path in life. Therefore, I did not know much about myself. I wanted to discover my strengths, weaknesses, desires, and what ultimately made me happy. The military was different, and it guaranteed a paycheck. Many of my attributes and life skills were learned in the U.S. Navy. A major part of my identity is based on my six years of military experience.

Chapter 2- Learning Who I Was

My recruiter picked me up on the night of July 4, 2000. I was eager and had my bags packed. Hours prior to his arrival, I made my rounds visiting family and friends. We took pictures, gave hugs, and our farewells. Around 11:00 p.m., I rode in the car with my recruiter to Charlotte, North Carolina. I gazed out the window and toward the stars wondering what life and fate had in store for me. As my eyes were fixed on the bright lights of the city, I thought of the endless possibilities of my military career. My future was a mystery, and I was content with figuring it out. Success at this moment was leaving my rural hometown and seeing different places and people. I felt like I had nothing to lose and did not bother looking back. Whatever the military had for me, I was going to take it and embrace it.

I spent the night at a hotel with other guys who were also entering the military. Instead of hanging out with new acquaintances and bouncing from room to room, I went to bed since it had been a long, tiring day. After waking up early the next morning, I quickly got dressed and walked expeditiously to the van, which took us to the Military Entrance Processing Station (MEPS). We spent the entire day filling out paperwork and getting medical examinations. Upon completion, we were sworn into the military and given orders to basic training.

They issued me a plane ticket to Chicago O'Hare, which was my first time flying on a commercial airliner. I knew there was another life waiting for me at my destination. I just took a leap of faith and prayed that I was going in the right direction. My heart throbbed as we taxied the runway, and I felt the plane lift into the air. This was the point of no return and my first time ever leaving my parents. After arriving at the airport, we waited a couple of hours for more flights to arrive. My anxiety level increased as I saw more people enter the room. We all eventually boarded a touring bus and headed to Great Lakes, Illinois.

Basic training was exactly what I expected. It was mentally and physically demanding. We met people from around the country as we were placed in divisions and assigned Recruit Division Commanders (RDC's). Other recruits, who started out as strangers, became friends and eventually brothers and sisters. Each week of training was comprised of team building activities and exercises. We also attended classes, and our RDC's ensured that we remained disciplined. After a couple of weeks, we sort of knew what to expect and how to stay in compliance. Even within our division, we had cliques and favorite friends just like in high school.

The weeks rolled by, and I looked forward to writing letters home since they were my only connection with the outside world. The only thing that was more exciting was receiving letters! I had already decided I wanted to work around aircraft, so I planned to be an airman. As the last weeks of training progressed, it saddened me to learn that my mother and girlfriend could not attend my basic training graduation. Despite the fact, I continued pressing on and motivating myself. Graduating basic training was still a remarkable experience after all the marching, reciting, tests, and physical training. Earning the privilege to wear that white uniform was well worth it. I proceeded to call my mother and girlfriend after the ceremony and enjoyed some liberty at the Navy Exchange.

Even after graduation, we remained on guard and maintained military bearing. Nothing would have been worse than getting set back in training for what was perceived as inappropriate behavior. We still needed to arrive at the airport and to our next duty station without upsetting an RDC. My biggest desire was to catch up on sleep. Sleep sounded like paradise after eight grueling weeks of physical and mental exhaustion. Words could not explain how I felt sitting on a touring bus, exiting the base, and heading to the airport. My next destination was Pensacola, Florida. I went from bitter cold weather in Great Lakes to smoldering heat in the south within a couple hours in September. Airman training was only two weeks long, and I had an opportunity to exercise some leadership abilities. The base was far more relaxed, and I enjoyed the downtime. Although I was not permitted to leave base, I found comfort in being able to call home more often and being an individual. Learning about aircraft carriers, aircraft, and ship life was exhilarating!

After two weeks of airman training, I was able to go home on two weeks of leave or vacation before reporting for duty on my ship. My time in Florida had been a pleasant adjustment. Ideally, I wanted to leave base, drive, and go to the beach. Therefore, I had been looking forward to returning home so I could have total freedom. After jumping on a flight and arriving back in Charlotte, I truly appreciated Lancaster, South Carolina. My hometown is small and has a slow-paced feel. Seeing my mother and girlfriend after a ten-week absence was a treasured moment. I enjoyed the simple things in life such as eating together, going to the mall, and relaxing at home.

Learning my personality was not easy. I did not know what I liked or disliked. I simply knew what I learned in my hometown. Any small town has its unique way of doing things. This was a chance for me to discover myself and my capabilities. Basic training taught me, foremost, that I must be a follower before I could become a leader. It also taught me to be detailed oriented, a team player, resourceful, respectful, and consistent. Showing honor in my work carries a tremendous amount of weight, even when others are not watching. Airman training taught me how to acquire leadership qualities. It takes more than knowing how to perform a job or task. I had to earn my peers' respect by taking on more responsibility and being accountable, no matter what the end result. This was tough for me because it is always easy to point fingers when something goes wrong. Being a leader meant engaging with the people around me to get an assignment done. I learned I must have social skills to be a leader. Establishing a bond, earning trust among my team, and fostering reliance was necessary to achieving our goals. This experience helped me gauge my leadership qualities, and I wanted to grow in this area.

Chapter 3- Joining the Workforce

I spent my two weeks of leave maximizing time with family and friends. The time had come for me to report to my ship. I booked a flight to Norfolk, Virginia where I caught an international flight to Bahrain. This was an enduring and adventurous experience, as it was my first time ever leaving the United States.

My overseas flight landed in the Azores, Italy, and Bahrain. Bahrain was extremely hot, and sweat beads rolled off my forehead while exiting the plane. My shipmates and I were placed in a hotel for the night and transported to base the next morning. We awaited a cargo plane, referred to as a COD, which flew us to our deployed ship in the Persian Gulf. The flight was bumpy but fun as we approached the ship. Flight personnel alerted us to be aware of the landing. On a commercial flight, landing at an airport is smooth and soft. This landing felt like we slammed onto the flight deck, and the plane came to an abrupt stop. A few minutes later, the cargo door opened, and I was welcomed onto USS *Abraham Lincoln* (CVN-72).

We deplaned and were guided into the ship, which looked nothing like I had ever imagined. The hallways and passageways were small and narrow. At this point, my buddies and I who flew across the Atlantic Ocean separated. I spent some time sitting at the Air Department office where I was assigned to a division. A representative came to retrieve my belongings and me. He escorted me to my new berthing or living space. I found a bunk and began stowing away my uniforms and gear. We went to the mess decks where we ate dinner. I was famished and had not eaten since we were on the flight across the Atlantic. Seeing the normal operations of the ship was astounding. Everyone was in working uniforms, usually dungarees or overalls, calling each other "shipmate." It was not the glamorization I had seen portrayed on Navy commercials at home. I thought Navy personnel routinely wore summer and dress whites. Instead, I discovered that upon deployment, they wear uniforms designed for utility and durability.

By the fourth day on the ship, reality began to set in. We worked twelve-hour shifts, which were tiresome. I missed my family. I can remember sitting on the floor beside my coffin rack getting teary eyed. I wondered if the next four years of my life was going to be this hectic.

I was assigned to the hangar bay where we repositioned and lifted aircraft on the elevators to the flight deck. It was tough keeping my uniform clean. I always looked forward to mealtime. Not only was it a chance to eat, but I enjoyed a break. It was so hot in the Persian Gulf at sea. We wore sleeves for protection and were encouraged to hydrate frequently. I began to look like a grease monkey because it was dirty and oily all over the hangar bay.

Eating lunch one day, I saw a higher ranking superior in khakis. He was clean and looked quite comfortable as he was passing through the mess decks. I immediately asked someone close to me if they knew his rank. A shipmate informed me that he was an officer. I proceeded asking questions about how I could become an officer. He told me I first had to go to college and get a degree. I was shocked! I thought all military personnel started as enlisted and worked their way through the ranks. **I began to think about going to college for the first time in my life.** This was a pivotal point in my life, as I viewed earning a degree as impossible or very difficult.

As the weeks passed, I enrolled in classes offered on the ship. I began attending one class at a time after I had completed my shift. It was not as difficult as I had imagined, and I found it enlightening to engage with shipmates over non-military topics. There was a social atmosphere in the classroom that allowed my mind to escape. I had no clue where it would lead, but I felt energized trying something new. It was a fun experience, and I was academically bettering myself. Who knew college classes could be so rewarding? I took two classes while I was out at sea and on deployment.

After we returned to Everett, Washington, I prided myself in continuing classes on base. Subconsciously, I had planted seeds for a lifetime. Attending these classes raised my self-esteem and gave me a chance to socialize with other intellectual students. The only challenge was that the ship would not stay docked in one location for long periods of time. It was my home; when it left, I had to go with it. This made my attendance sporadic as I tried maintaining momentum with my classes.

The USS *Abraham Lincoln* (CVN-72) eventually ported in Bremerton where it underwent a six-month overhaul. We were assigned living quarters, which looked similar to college dorm rooms. Best of all, I only had one roommate who was really easy to get along with. I racked my brain on how I could attend college on a regular basis.

Working on the ship during the overhaul was tedious and busy due to many projects going on simultaneously. I began seeing enlisted personnel walking around who were clean and kempt. I approached one of them and asked where he worked. This shipmate was a dental technician. He was articulate and carried himself with high esteem. I was a bit tired of being a grease monkey in the Air Department. I wondered what it would take to become a dental technician, so I spoke with my departmental personnel who seemed to discourage the idea of anyone transitioning into a new field. I began doing my own in-depth research, discovering the criteria and protocol for cross-training. It was going to be a tough ride, but I was determined to make it happen.

First of all, I had to accumulate forty hours of on-the-job training in the dental field if I wanted to become a technician. Then, I would have to request a departmental transfer and work for several months until I was sent to dental technician school. There were limited students accepted, and this school was eventually shutting down due to an upcoming medical/dental merger. Thirdly, I had to impress a board of high-ranking enlisted personnel to prove I was worthy of the investment. I was up for the challenge so I could work in a clean, professional environment.

I became well acquainted with my ship's dental personnel. I wanted to know everything about the dental field from administration, x-rays, and routine procedures to complex procedures. I found it beneficial to be a great resource for the department because they could ultimately change my quality of life forever. Showing enthusiasm for the techs, chiefs and dentists made a lasting impression. However, I encountered a slight problem. Dental offices were open during normal business hours, which were 8 a.m.-5 p.m. I also had to work that same time frame in the Air Department. How could I work in two departments at the same time? After a few days of racking my brain, I determined that sacrificing my lunch hour to accumulate dental OJT was the solution. I thought gradual progress was better than no progress. My lunch hour was the only time during the workday that my higher-ups could not control. It would be a huge sacrifice because I enjoyed my lunch break and winding down midday.

After a week of OJT in the dental department, my hours looked miniscule, but I was determined to stick with it. After a couple of weeks, I began to see slight progress. I was not able to attend OJT every day due to my demanding job as an airman. After two grueling months, I had finally accumulated forty hours of OJT. During this time, I built relationships with new shipmates, proved my commitment to the team, and convinced myself that I could achieve anything I put my mind to.

At this point, I was able to request a departmental transfer, and it felt marvelous! After several months, a board of chiefs approved my request, and my life began to slowly change for the better.

After transferring to the dental department, my quality of life increased drastically. However, I faced new challenges. The expectation was for me to perform like a full-fledged dental technician. I was still an airman awaiting orders to dental technician school. Therefore, I did not have a lot of book knowledge in this new field. My shipmates explained things to me and showed me everything I needed to know. I learned what true camaraderie was after twenty years of living on earth. I conducted research on where I would like to transfer upon completion of dental technician school.

Attending college was still my priority as I transitioned into my new profession. I wanted to be stationed close to home where I could be close to my girlfriend, family, and friends. I knew there were only Army, Air Force and Marine bases in South Carolina. The closest, sizable bases that I could realistically transfer to were in Norfolk, Virginia and Jacksonville, Florida. It would be a stretch going home on the weekends though, and neither of these was my ideal traveling distance.

After searching all military bases in South Carolina, I thought more closely about Marine Corps Recruit Depot Parris Island. It was even more interesting that dental technicians worked on that base! I discovered that Navy personnel provided all medical, dental, and religious support for the Marines. Better yet, there was a university located in Beaufort, South Carolina where I could attend college classes. How wonderful would it be to work during the day and go to class during the evenings! The University of South Carolina Beaufort seemed like a small school when visiting their website. Looking at the big picture, I could potentially transfer to a geographically desired area and complete my degree; at least that was the plan. I felt the "yes factor," which gave me the green light to proceed with my strategic plan.

After nine months of working in the dental field, the time came to depart USS *Abraham Lincoln* (CVN-72). It was

October, and the ship had deployed in July. In retrospect as we were getting underway, I stood from the flight deck overlooking the base. We manned the rails as family and friends waved giving their farewells. I stared at Mount Rainier as we slowly sailed away and did not expect to miss Everett, Washington.

During my last few days on the ship, I was overwhelmed with my exciting new adventure. I gave farewells to my shipmates and packed my belongings for a flight off the ship. I was one of few sailors who flew onto an aircraft carrier upon arrival and off an aircraft carrier upon departure of a duty station. Taking off from the flight deck was exhilarating and beyond imaginable. I had dreamed about this moment for at least a year. The flight to Bahrain was bumpy again but worth it. After arriving to the base, I stepped on land, which felt amazing. I stretched my arms out in the air and said "Civilization!" I had a sense of accomplishment and new goals in the military. I had faith and hope to keep moving forward in my career.

We waited at the base until a commercial military plane flew us to Italy and directly back to Norfolk, Virginia. I stayed in a hotel after arriving in Virginia. That next morning, I caught a ride to the Amtrak station and bought a ticket to Charlotte, North Carolina. It was a long, comforting ride down to the Carolinas. My mother picked me up from the train station, and we proceeded home to Lancaster, South Carolina.

I was eager to buy a car since I was going to be stationed on shore duty. My mother and I drove to Charlotte's Independence Blvd, which is a playground for a car buying prospect. We ended up at the Infiniti dealership. We browsed their inventory, and I found a silver G20 sedan with a spoiler. I had wanted this car since high school. It looked sporty; and after test-driving it, I had to have it. After a couple of hours, I was approved for financing. I even took a photo with my salesperson since she was professional, straightforward, and a pleasure to work with. It felt sensational driving my new car home.

I had about two weeks of leave before reporting to Sheppard Air Force base in Wichita Falls, Texas. During this time, I enjoyed my time off at home by visiting and catching up with family. They were so proud of me, and I was even prouder of myself.

I stopped in Spartanburg, South Carolina to visit my girlfriend for the weekend. The journey began, as this was my first time driving cross-country. It was something I had always wanted to do. I drove to Atlanta where I got stuck in traffic; then I proceeded on Interstate 20 toward Birmingham, Alabama where I stayed the night. I had about three days to travel and enjoyed each city by exploring their shopping malls. The next morning, I set out for Jackson, Mississippi. I took pride in crossing the Mississippi River since it was my first time driving into a different time zone. After a few hours, I ended up in Shreveport, Louisiana and stayed the night. The next day was going to be a big day. I was looking forward to Texas and dental technician school.

After a good night's sleep, I saddled up the next morning. There is something I love about early morning driving. Maybe it is the cool, crisp air or just knowing that I had a productive day ahead of me. Going on road trips 30 minutes before dawn is the best. I beat the morning rush hour and get to see the bright orange sun as it peeks over the horizon.

After crossing the Texas/Louisiana state line, I found it true that everything is bigger in the Lone Star State. I was driving on a long, desolate road and saw a tumbleweed bounce across the road in the wind. Until that moment, I had only seen that in movies. It seemed like it took forever driving through Texas since there was little to no scenery on the country roads. Maybe I was just tired after my three-day long drive. I finally arrived in Dallas and was blown away by the beauty of the city. I still had another two hours until I arrived in Wichita Falls. I passed the time by listening to CDs. Then lo and behold, I began to see signs for Sheppard Air Force Base. My adrenaline was pumping when I arrived to my new base, school, and home for the next three months.

When I checked in with Navy personnel, they greeted me with open arms. They assigned me living quarters, and I was eager to explore the base. I quickly noticed my cell phone signal was weak throughout the base. I knew my time there was going to be short-lived, so I did not let it bother me. It was exciting meeting new shipmates and airmen. The first day of class was awesome, and my instructors were amazing. I was immediately placed in a leadership role since I was a fleet returnee. I was responsible for mustering my classmates and getting everyone to class on time. Since I was more experienced than my classmates, they respected me and followed my lead. Most of the classroom teaching, I had already experienced on the ship. It was just as important to attain the book knowledge and pass structured tests. I realized even though the subject matter was in-depth, I would only absorb as much as I put my mind to. These learning tools were necessary for college and other disciplines in life. So, weeks passed, and I led my class through training. They actually taught me people skills, and in return I taught them Navy ship life philosophy. They always asked me numerous questions about my ship, and I prided myself in providing realistic scenarios for them.

After class, I would occasionally head into town to eat. I also drove to Midwestern State University to briefly get away and observe college students. I managed to use their computer lab to check my email. I hoped to make some friends on campus, but the students usually kept to themselves. It felt great being able to jump in my car and go. I even drove up to Lawton, Oklahoma and found a small mall where I saw other military personnel.

After a few months of molding my dental classmates and learning dental technician basics, graduation was approaching. I looked forward to getting settled into my permanent duty station. The weather began to cool off, and Texas wasn't so hot anymore. When determining my next duty station, my instructors were shocked when I requested a transfer to Parris Island, South Carolina. They were expecting me to ask for San Diego, Hawaii, Japan, or Italy on my wish list. These were considered the sought out and exotic bases for Navy personnel. However, I remained focused on college and the big picture. They granted my request although they did not understand my ultimate plan of being close to home. Graduation was an inspiring day, and I was itching to get on the road to arrive in South Carolina a couple days before Christmas.

I packed up the car and made my way to the interstate. I could finally go back to South Carolina and live what I thought was a normal life. I did not spend a lot of time stopping as wintry weather had swept across the South. There were numerous car accidents from Texas to Georgia. I remember counting at least twenty collisions related to rain, sleet, and icy roads. Getting home safely and timely was my main objective. I felt joyous crossing into South Carolina and only had a couple more hours until I would arrive at home. I was so eager to see my family. When arriving back to Lancaster, they greeted me with big hugs and kisses.

Joining the workforce was an eye-opening experience. Working full-time was not what I perceived as a teenager. I thought it was as simple as getting up and performing a job each day. However, it was much more. Being in the work force meant being a part of a team and believing in something bigger than myself. We all strive to be a part of teams within our workplaces, neighborhoods, and communities. Being a law-abiding citizen who looks out for the welfare others is important. The economy works a lot better when able-bodied adults and age-appropriate teenagers are fully employed. Whether someone has a GED and works in a fast food restaurant or an MBA from an Ivy League school, their participation in the work force is necessary to keep a growing and progressive economy. I learned a new set of life skills, such as importance of creating and maintaining healthy relationships with coworkers, paying taxes, and other financial obligations. The ship life taught me what it is like to work enduring shifts. I learned how to think strategically to get what I want in life. Life can be played like a game of checkers, but it should be played strategically like chess. Careful thinking, clever moves, and clear plans will propel your life into what you want it to be. Leaving my ship and transitioning into shore duty benefited me immensely. It fortified my confidence and gave me the drive to move forward. I found myself setting new goals when I arrived at my new duty station, which is what the work force should do if you want to move forward. Challenging yourself to achieve new goals in the workplace is what helps you move up to the next level in life.

Chapter 4- Starting College

I enjoyed a week and a half at home before reporting to my new permanent duty station. I felt blessed to spend Christmas at home with my family. My shipmates were still deployed and out at sea. It was only a three-hour drive to Beaufort, South Carolina. My previous duty station was a five-hour flight from my hometown. I arrived late, and visibility was limited due to nightfall. I was hoping to see palm trees since Everett, Washington usually offered cloudy, gloomy weather. I checked into a hotel that night and mentally prepared myself for an all-new military experience.

The next morning, I dressed in my uniform and drove to base. After arriving to the dental clinic and scoping out my home for the next three years, I began to see new faces. They introduced me during morning roll and I began the check-in process. Working at the Naval Dental Center (NDC) was a breeze compared to ship life. Wearing comfortable scrubs and sitting in an air-conditioned office trumped being in long sleeves floating in the Persian Gulf heat! Working 7 a.m.-4 p.m. was a light shift, and I found myself bored after getting off work during the first couple weeks. I found a part-time job at a department clothing store to make use of my time before starting school. The semester had already begun at University South Carolina Beaufort, and I needed time to get acclimated to my new surroundings. I had not made any new friends yet, so the part-time job gave me something to do with my time after leaving the dental clinic.

After four months, I enrolled in USCB and began taking my first class on base. They had a program, which allowed military personnel to attend class after work. When speaking with a guidance counselor, I was a bit overwhelmed at the curriculum and how many classes I had to complete. Fortunately, my classes from the ship transferred over to my new school. I wanted my degree as soon as possible and realized quickly that I needed to adjust my schedule. Working during the day placed time constraints on how many classes I could take in the evenings. I began taking one eight-week course at a time. **My first class was fun, and I learned to just enjoy the experience instead of rushing to absorb the material.** Interacting with my professor and classmates made me feel socially accepted in the civilian sector. As I began taking more classes, they were offered on campus, at the United States Marine Corps (USMC) Air Station, or in Historical Downtown Beaufort on campus. I developed an academic routine and picked up momentum. My military coworkers and friends invited me places, but I was really busy with school and chose to stay intellectually disciplined.

After several months, I began training as an Expanded Functions Technician. In this capacity, I took on more responsibility with patients. I learned to clean teeth, take precise x-rays, and fill cavities with silver amalgam. I became an above-average technician and began to receive recognition from dental officers. I eventually transferred to the Air Station Dental Clinic where I was expected to take on more of a leadership role. This was a small clinic, which provided dental health to about 3,000 patients. The workload certainly increased as I transitioned from assisting and front desk administration to the x-ray department. I was also assigned auxiliary duties such a Central Sterilization Room (CSR) Technician and Material Management Coordinator (ordering supplies). I eventually picked up rank when I passed the Dental Technician Third Class test. After becoming a Third Class Petty Officer, my superiors expected more professional development. I began training personnel in the areas of x-ray, CSR, and hazardous material duties. It was so much to do every day that I rarely finished my daily "to do" list. Dental officers needed specialized equipment, patients needed a variety of treatment plans, and I was the guy who facilitated and delivered.

We also began mandatory physical training after work, which threw a wrench into my school schedule. My superiors cared more about my commitment to the Navy. Anything else outside of my military duties came second. I became overwhelmed and increasingly frustrated, as I found less time to focus on my studies between patients and consolidating work tasks. I began working more efficiently, and my efforts were noticed. In December of 2004, I earned my Associate Degree in Science at USCB and was recognized by my direct supervisors.

After receiving my two-year degree, I realized that college was not as difficult as I always imagined. It is also not a race. I had to be committed and dedicated to it just like anything else. As long as I invested my time wisely and developed progressive patterns, I would succeed. The new problem I encountered was the time of day my new business courses were offered. Most of them were available during the day while I was working. I had another challenge to overcome! I submitted a formal request to attend class during the afternoon, and it was denied. If they approved my request, they would have to approve others, thus affecting manpower. I understood and did not take it to heart. Strategic planning became critical, and I signed up well in advance for business classes offered in the evenings. I even took summer courses to expedite my progress.

The Navy extended twelve hours or four classes of tuition assistance each year, and I used it diligently. To get tuition assistance, a sailor would have to be above average. His work performance had to be outstanding, and he had to pass his physical fitness test every six months. These two criteria were factored heavily into whether a sailor would receive financial assistance and ultimately earn a degree.

I was happy with my life and the direction it was going. I knew my job well, and I was making progress toward my degree. I still struggled with making and keeping friends. Since work and school took up so much of my time, it was tough trying to stay committed to my relationship and friendships. The long distance relationship took an emotional toll and my girlfriend and I eventually broke up. All my acquaintances had their own lives, and I found it easier to be more introverted. I spent my time shopping and eating at restaurants to pass the time on the weekends. Retail therapy simply gave me something to do instead of staying home bored and lonely. I was in my 20's and very materialistic with cars, electronics, and clothes. I upgraded my Infiniti to a BMW 3 Series, bought expensive devices (such as digital cameras and a satellite radio), and shopped for clothes at Banana Republic. I saw no harm in my purchase habits because it kept me busy and out of trouble. The only problem was when cash was low, I would charge these items on credit cards. I always made on-time, monthly payments and had a credit score over 700. My taste became more expensive over the years, and I began to charge more because I thought it helped my credit to pay down high balances. I even placed my mother on my credit card account and allowed her to use it once a week for filling up her gas tank. My rationale was that I would finish my time-in-service, complete my degree, and get a higher paying job to pay off all the debt I had incurred.

My three years at Parris Island and MCAS Beaufort slowly came to a close. I had less than a year left on active duty and had to make a crucial decision to reenlist in the Navy or become a civilian again. If I continued serving in the military, it was mandatory to change duty stations. I had worked so hard to complete my bachelor's degree and did not want to relocate. I only had six semesters to go. I was also going to be deployed to Iraq with my fellow Marines if I reenlisted. I made a careful, conscious decision to complete my six-year time-in-service and pursue college full-time. My chain of command seemed disappointed in my decision to leave the military. I enjoyed my military experience, but my frame of mind was rooted into finishing my bachelor's degree and starting a business career.

Transitioning from military to civilian life was not as easy as I expected. I had begun preparing several months in advance before my contract was complete. Simple responsibilities such as living on a stricter budget, paying down bills, and looking for a place to live and work became my priorities. I was finally going to be able to use the Montgomery GI Bill (MGIB) while in college. Many young veterans neglect taking full advantage of their military educational benefits. I chose to live on campus at the Bluffton University dorms since it was new and within walking distance from class. After completing my medical exams for separation from the Navy, I requested a month of leave prior to my separation date, which was approved. My dental shipmates hosted a luncheon for me during my last week and gave farewells for my new life endeavor. Moving into the new apartment on campus felt like a dream!

Starting college felt like a success. In retrospect, I wish that I had started college immediately after high school. It was a lot simpler and more engaging than I expected. I was in a rush to make money after high school and was not focused on a long-term goal. However, it took me two years to get on track. Optimistically speaking, it was better for me to join the work force and see how competitive it was so that I could take college seriously. I began learning more about myself as I worked full-time and participated in evening classes. I became more disciplined as a person and relied less on others telling me what to do and how to do it. Taking the initiative in my life put me in the cockpit of achieving my dreams. Making choices can be scary and intimidating for teenagers and adults. However, one good choice leads to another, and over time, those choices add up. Beginning my college experience one class at a time helped me learn time management skills as well as my strengths and weaknesses. A couple of my weaknesses were loneliness and a lack of frugality. As a result, I spent my time buying a lot of stuff I did not need to fill the empty void of not having family around. Many active duty members experience loneliness, and it is important to build relationships with people so that you can support each other. Looking back ten years, I could have gone without all those electronics and clothes. The electronic toys became obsolete, and I outgrew or stained the clothes. The bottom line is I do not have those things anymore, and I financially overextended myself acquiring those unnecessary items. I could have also gone without an Infiniti or BMW. A used Honda or Nissan would have worked just as well and would have cost a fraction of what I paid for those luxury cars. As a young adult, buying those items gave me a sense of status and style. It made me feel important and like I was doing something right. It is ok to have nice things, but it is more important to buy necessities first and luxuries last. Starting college gave me the tools to prioritize daily tasks, maximize resources, and become more financially aware of my long-term future.

Chapter 5- Learning to Be a Student

I was overwhelmed with joy, excitement, and enthusiasm with my new course of civilian life. The atmospherics of the school and apartment were refreshing. I felt like God had given me a new chance at life. Everything I had worked so hard for on USS *Abraham Lincoln* (CVN-72) had come into fruition. It was my deepest desire to be a full-time student and live on campus. I felt this college experience was going to be great for learning social skills and more professionalism. I was exposed to like-minded, intellectual individuals who also wanted degrees. I quickly learned to network especially when resources, such as time and money, were limited.

Sleeping in during the mornings felt strange. I always woke up at 5:00 or 6:00 in the morning and could not go back to sleep. The military programmed me to get up at this time. Class usually started by 9:00 or 10:00 a.m. My mornings were consumed with working out in the gym and jogging. Exercise was the only thing that felt normal because everything else had changed. I was 24 years old and a little bit older than most of my classmates, most of whom were 19-21 years old. In an attempt to fit in, I often did not reveal my age. People treated me differently after discovering I had graduated high school in 2000, served six years in the military, and traveled all over the world. I was not ashamed of my life or experiences but I wanted my peers to perceive me as an aspiring student.

I picked up a part-time job at a car dealership across the street as a service department driver. Before, between, and after classes, I would deliver loaner vehicles to customers in the surrounding area. I also began working at an electronic store on Hilton Head Island. It was a retail job that guaranteed hours each week. My third job was being a substitute at Hilton Head Island High School. I enjoyed being a role model for teenagers, but this job was short-lived because I could only substitute on Fridays since I had classes Monday through Thursday.

I thought having three part-time jobs along with my military educational benefits would be enough to keep me financially stable. However, it was hardly enough to cover basic expenses. The hours at each job were so inconsistent and the enrollment for benefits took longer than expected. **It was mid-September 2006, and I began to have financial woes for the first time in my life.** I knew it was a problem when I began receiving collection calls for my car and credit cards. I was in serious trouble! My only option was to work more hours, which was tough because I had to attend classes, study, and do homework. I managed to detail cars every once in a while, which really helped.

Soon, however, I began to get behind on my rent, which was a whopping $600 a month. I knew that I was not going to be able to afford rent, a $500 car payment, insurance, gas, and groceries. I remained prayerful, hoping that my financial distress would end soon. I had some life-changing decisions to make. Even after receiving my military benefits, I was drowning in debt and overextended. Those funds were supposed to last an entire month. By October, I knew that I needed to make a sacrifice. It hurt deeply, but I had to voluntarily surrender the BMW to my auto lender. It became too expensive, and paying for premium gas was almost impossible. It took about a month to set up an arrangement for returning the vehicle. Instead of having a tow truck show up, I chose to drive the car to Charlotte, North Carolina where I turned it over to a designated repossession company. By December, the car needed expensive maintenance and a set of new tires that I could not afford. Although I missed the car, it was a huge financial relief getting rid of it.

As Christmas 2006 approached, I was almost finished with my first semester of college. The previous few months had been chaotic and filled with a bombardment of distractions. I irrationally decided when to study and do homework, when to work and spend time with friends. As a result, my grades suffered. That may be an understatement.

My grades were horrible, and I even failed Art History. I asked myself on many occasions how this happened. As long as I was active duty and going to class, I did exceptionally well. As soon as I pursued school full-time, I performed poorly. Maybe I had overloaded myself with too many classes and jobs. I had lost focus on my ultimate goal. I took a huge pay cut separating from the Navy to earn my degree and I was off to a terrible start.

Thoughts of recourse flooded my mind. How could I maintain my life and complete school successfully? So far, I had ruined my credit score, lost my car, and committed academic suicide. **Had college become a curse?** I was determined the next semester would be better. I pledged to be more resourceful and practice better time management.

I enrolled in my Spring 2007 classes with a go-getter mentality. I focused more on reading my lessons and doing homework in a quieter environment. I discovered going to the campus library at 8 a.m. was helpful. There were many computers available and few people to distract me from studies. Later around 10 a.m. is when students began showing up and socializing. Those two hours of isolation allowed me to absorb a considerable amount of study material. The rest of the day was devoted to class, work, and social outings. I found a stable and reliable front desk job at a hotel in Bluffton. This position allowed me to work on my studies when business was slow. I had access to a computer, a countertop to lay out books, and an air conditioned, quiet office space. I left the retail job and spent less time driving at the dealership. My academic success was priority this semester, and I wanted to cross the finish line and earn my degree.

    I placed myself on an even stricter time management schedule the following semesters. Devoting my time to reading, studying, homework, and group projects became crucial. Hanging out with friends and going to the beach became less important. I just wanted to finish college so I could find a job and begin working. My days continued starting early with a workout, breakfast, and going to the library. I was usually exhausted but kept my eye on the prize. After a year of living on campus, some friends and I found a house to rent in Hardeeville. We were all eager to economize, so we shared a dwelling and resources so we could graduate. My new study strategy and budgetary limitations seemed to work in my favor. Semester after semester, I diligently completed courses above average.

I reached my last semester and loaded my schedule with classes. One of the classes was calculus. This was my second time taking calculus, and I needed quite a bit of help in this area. The first time was during the summer of 2004 while I was on active duty, and it was an eight-week accelerated course. After working during the day, I just did not have the brainpower to wrap my mind around all the formulas and methodologies. I wanted this second time to be a game changer. Most math teachers and professors did not spoon feed this advanced material to me. I learned that I must take initiative by spending more time studying and doing homework. Finding an affordable tutor would have helped tremendously. I did not find a tutor this time because I was convinced and a bit naive that I could do it myself. I also did not have the money to pay a tutor, and my time was limited. This class was different because it was sixteen weeks long.

After the first month, my grades began slipping. The lessons were cumulative, so if you reached a level that did not make sense, the rest of the semester would be challenging. I spent more time studying and doing homework but just did not absorb the material. After bombing a couple of tests, I hoped that my professor would have mercy on me and just pass me with a C. I spent the entire semester being an overachiever and this was my only weak area.

My parents and aunt made plans to drive down to see me graduate. This milestone meant a lot to me. The thrill and intensity increased each week. I was looking forward to creating a memorable experience with my family. When filling out my cap and gown form, I could see the light at the end of the tunnel. I spoke with my general manager and accommodated my family with rooms at the hotel where I worked.

The day of graduation came, and the weather was beautiful. I was nervous, happy, and a little sad all at the same time. Life was about to change AGAIN. My classmates and I looked great, and I took pride in walking across the stage with them. I will never forget the moment I grabbed my scroll and shook hands with the dean. I was ecstatic and had the biggest smile on my face! I had officially graduated high school, honorably served six years in the Navy, and graduated college.

After the ceremony, my family and I went to a seafood restaurant and celebrated with a wonderful meal. The next day, there was an up close photo of my face on the front page of the *Bluffton Today* announcing the USCB graduation. I felt accomplished and relieved.

A few days later, my spring semester grades were released, and I found out that I had failed calculus again. I was not surprised, and my performance had been subpar. Although I had walked across the stage during graduation, I would not receive a degree until I took calculus for a third time and passed it. I wanted to quit and just move on with life. What would it matter anyway? I had attended college and lived the experience. I had taken other business courses and passed with flying colors. Maybe I could just pretend that I passed calculus and graduated. After careful consideration, I decided that I did not want my academia to end this way. Calculus was offered again during the spring semester of next school year. So, I was trapped living in Hardeeville working at a low-paying job when I longed to be in Charlotte starting a new business career. I remained prayerful and kept reassuring myself.

I continued living at the house with the guys and found a data entry position in Savannah, Georgia with a job placement agency. For the first time, I had the privilege of working in a shirt and tie. I loved dressing up and going to work. I also found a part-time job at another hotel. This time, I was working at a restaurant within the hotel as a bartender and server. So, I worked my office job from 8 a.m.-5 p.m. and my bartending job from 5:30-11:30 p.m. Both of those jobs were located in the Historic District of Savannah. These opportunities gave me a chance to be a part of a small city metro area. I networked in hopes of finding a full-time, permanent position with benefits. My position as a data entry and mail clerk at a shipping company was short-lived, as my assignment was complete after three months. I went back to my dealership to drive but continued working as a bartender. I just wanted the spring season to hurry up so I could take calculus class and move on with life.

I had to pay for this class out of pocket. Since I was no longer enrolled at least half of the time, I could not apply for a student loan. It was a Catch-22. I could not get a loan and made too little money to pay tuition myself. Thank goodness for my ex-girlfriend's mother who was like a mother to me. Even though her daughter and I split up, she remained in touch and helped me cover the tuition. I only had to pay for the calculus book. The unwelcomed news was the class being offered only at the North Beaufort Campus. So, I had to drive 20-25 minutes to and from class which created a gas expense. Calculus seemed to have plagued my life.

When the spring registration approached, I eagerly signed up for the class. I would have been mortified if the class filled up before I enrolled. It was offered during the evenings two days a week. I began working solely at the dealership and discontinued my bartending job. I kept busy traveling to Hilton Head, Beaufort, and Savannah delivering vehicles. I prided myself in working overtime as it became available. I drove so much that my eyes were sore and red by the end of the day.

Once class was in session, I could not believe I was studying the same material for the third time. I was so tired of looking at calculus and was determined to get it right this time, especially since it was my last class. I just had to plan ahead and be prepared. After driving all day, it was exhausting having to commute to class. I was tired of complaining and just wanted to get it done. This class was an obstacle for my professional development. I paid close attention, asked a lot of questions, and had a classmate who held me accountable. Stephen and I sat next to each other in class and raved about how beautiful our young professor was. So, I had motivation from him and her to keep me going. As the weeks progressed, work began to interfere with going to class. One day, I was expected to make a delivery to a customer when it cut into my class time. It was a lot of pressure. If I did not accept the work, they may find others to help and slowly phase me out. If I chose to attend class, I eventually would not be able to pay bills. My finances were tight, and I could not afford to lose this job. However, I could not fail this class again either. My GPA had suffered enough! I began reminding the service advisors on Tuesday and Thursday mornings that I was available until 5 p.m. If they needed a vehicle delivered during the evening, they had to call someone else. It made me cringe to give away hours, but I was willing to make sacrifices in an effort to pass calculus and move onto a career.

    I consistently studied and did my homework. I even arrived an hour early some days for tutoring. I showed my professor that I was dedicated to learning the material. Showing enthusiasm convinced my professor that I was not just seeking a passing grade. I maintained a satisfactory grade throughout the semester, and it was time for finals. I was fortunately still employed at the dealership. The thought of being replaced was just an illusion. I devoted weeks, days, and hours of effort into absorbing all the material I learned during the semester for the final. Taking the test was nerve-racking, and I had so much at stake. After completing the final, I was not sure how I had performed. I had to wait about a week to view my grade online. The suspense was tantalizing.

When the final grade became available, I could not look at it. I signed into my student portal and had my coworkers look at my grade. When they started cheering, I immediately knew I passed. After three attempts, I passed calculus with a B.

Since the job market in the Lowcountry was not promising, I decided to finally move to Charlotte. My dream was to work uptown in one of the tall buildings and enjoy the fast, metropolitan life. I wanted to mingle with other professionals in shirt and tie. I submitted my two-week notice at the dealership after three years of employment and began preparing to move home with my mother. I really wanted my own apartment, but I had to find a job first. I also had to buy a car. This chapter of my life slowly came to a close, and it was time to start fresh. I took the Amtrak from Savannah to the Carolinas. My mother picked me up at the train station, and we headed home.

My idea of attending college was distorted as I transitioned from active duty military to civilian life. I exited the military while being financially overleveraged. All the shopping and material items I bought meant nothing as a college student trying to survive. If there were one thing I could have changed, I would have lived a more simple life when I had disposable income. Just because I had the necessary means to purchase items didn't mean I had to be a big spender.

When making purchases, ask yourself what that particular item will mean in the next five or ten years. Will you still have it, or will it be buried in a landfill? Practicing frugality has many benefits. Although you may look or appear basic, you will have purchasing power to care for your needs.

I learned more about myself when attending college full-time. I discovered my depth of ambition and desire to succeed. I pushed myself academically and in the workplace more than ever. I faced some obstacles along the way such as lack of income and distractions.

It is normal to get off track in life. When this happens, you must remind yourself of your ultimate goal. When you remember your objectives, it makes it easier to get back on track. I used to sweat the small stuff and lost focus of the big picture. It is ok to make mistakes. Mistakes are designed to help you understand you have gone in the wrong direction. I would question the validity of anyone who claims they have never made a mistake.

In hindsight, learning to be a student was tough. I had to reset the way I processed information. When I aligned my thoughts and actions, I was able to achieve my goal of graduating college. It was a lengthy, tedious process that ensued for six years. However, I learned just as much about myself in college as I learned about myself in the military.

When you are facing a difficult time, remember that adversity will always bring out the better person in you.

Chapter 6 – Post-College

The Great Recession struck the United States in 2008, and about 2.6 million Americans lost their jobs. Charlotte, North Carolina, the second largest banking center in the country, was adversely impacted. I read about economic downfalls in history books as a child but did not understand their depth of hardship. I had faith my new degree would set me apart from other job seekers. My goal was to find a job within two weeks. However, the job market in Charlotte was different from the one in Savannah. **To my disadvantage, employers were seeking candidates with degrees AND experience.** My only work experience was as an airman and dental technician in the Navy, in addition to retail and hospitality management in college. I also discovered there was a surplus of degreed candidates in this area.

I began applying for jobs feverishly. I updated my resume and made necessary changes. My mother lived about a mile from USC Lancaster. Since I was a USC Beaufort graduate, I could still use my login to access the student portal on the computers. I found myself at the computer lab searching and applying for jobs every day. The library was open mostly Monday through Thursday, and mornings were the best time to go. Since my mother used her car for work, I walked (and jogged some days) over to the school. After two solid weeks, the phone didn't ring once. I began to get weary. Bills were starting to pile up, and I needed to help my mother with household expenses. After three weeks of applying to jobs online and receiving no response, I considered a different approach. I began taking my mother to work and using the car to search for jobs in person. Time was of the essence, and I needed to start working ASAP. I visited hotels and malls with hopes of finding a simple position just to survive. I walked through the mall looking for "We're hiring" or "Help wanted" signs. I did not find many or even one for that matter.

So, I began asking questions to store personnel. I proceeded into an electronic store and remembered my experience working in retail. I felt a bit shy but mustered up enough courage and asked the sales associate if they were hiring. She told me they were having a job fair that same day. I knew God was on my side, as she gave me directions to the office. I felt a surge of excitement and drove over to the corporate office where I applied for a sales associate position. They interviewed me on the spot and said they would contact me.

It had been six days, so I gave them a call. The recruiter informed me she had planned on calling me that day. I had just beaten her to the punch. I started work on my birthday and did not mind. I was not thrilled about the $7.00 an hour wage, but it was a job; I was making progress. Maybe now I could network and meet some people who could get me in the door at a reputable company with decent pay.

I trained and learned how to use the register, educated myself on the phone plans, and familiarized myself with the store. That next week, a new associate started, and he was amicable. His name was Casey, and we had a lot in common. We were the same age, both ex-military, and looking for an apartment. He was staying with his sister and had enrolled in school. Casey and I became good friends, and he even trained me on how to navigate the computer when selling mobile phones. Our manager assigned us to a kiosk in the mall, and our job was to convince customers to switch mobile phone providers. This was, by far, not my dream job but I was being productive and no longer sitting at my mother's house.

Working in retail was tough. I had to juggle customer service and inventory management. My least favorite task was changing price tags for sales. There were hundreds of them every day. The workload was unbearable, as they scheduled us to work open to close almost every day. After receiving my paycheck, I felt shorted. As the weeks passed, the labor-intensive tasks seemed to outweigh each paycheck. There was absolutely no way I could afford an apartment or car on this pay.

The next month, I went with my dad when he bought a new car. He kept his expected trade-in and allowed me to drive it for work. The bug-looking car was in poor condition, but it changed my life. It gave me freedom and mobility. I no longer had to hog my mother's car and could be on my own schedule. I also began looking around for other jobs so I could get an apartment in Charlotte. The one-hour commute was tiresome, and my gas expenses consumed a considerable amount of my paycheck.

I interviewed for a front desk agent position at a nearby hotel. I was offered the job, but it was only part-time. After a month and a half in retail, I felt like I was spinning my wheels. Most of the money I earned was going toward purchasing fast food and gas. If I wanted my life to change, I once again had to take a different approach. I quit the job at the electronic store and accepted the offer at the hotel. The hotel paid more, and the experience would look better on my resume. My mother did not understand or appreciate my switching jobs and made it clear she wanted me to move out of her house. My living situation became very complex. I moved in with my dad as Casey and I began to look at apartments for rent. Casey and I were able to work as a team because I had a better credit history, and he was expecting a huge military educational disbursement. With my credit and his money, we could rent a two-bedroom apartment together.

After two weeks of staying with my dad, we found an apartment, and our application was approved. We just had to wait for Casey's direct deposit to post. So far, this Charlotte move was working out slowly but surely. My dad knew someone who was affiliated with a bank employee in Charlotte. I called him, and he instructed me to apply for a position through an employment agency. The employment agency interviewed me right away and thought I would be a good fit for a fraud specialist position. They forwarded my resume over to the bank's loss prevention department, and they also interviewed me. My interviewers were easygoing and made me feel welcome. I felt like I made a lasting impression and anticipated a callback.

While waiting to hear back from the bank, I continued working at the hotel. Casey finally received his direct deposit, and we signed our lease agreement for the apartment. I had many irons in the fire, and life was finally taking a favorable turn. We moved into our new apartment, and I felt relieved of my usual commute. It took about three weeks before my employment agency contacted me about the fraud position. My quality of life was increasing with the car, apartment, and a better paying job. This well-deserved transitional period was the beginning of a new season.

After starting training at the bank's call center, I continued to work at the hotel on my days off. It was exhausting, but I had to maintain cash flow and momentum. I met some awesome people working at the call center. I had finally accomplished one of several objectives in Charlotte. After a couple months, I became more efficient at my job and time management. Casey and I made great roommates, and life was moving forward.

I took pleasure in going to new restaurants and uptown to hang out and attend weekly jazz events. I enjoyed meeting new acquaintances and making connections. Wine tasting became a new hobby. The metro life consisted of wearing business casual clothes for work and going to my favorite happy hour spots. I even found a good deal on a Blackberry phone, which made me more productive while on the go. Barbequing and cooking became my hobbies and consumed my time at home. There was something unique about spices, seasoning, and marinades that kept me in the kitchen. It was fun prepping food and experimenting with different meats. Best of all, it was a healthier and cheaper option than eating out all the time.

After six months of working at the call center, I was eager to be hired as a permanent employee. There was a pending merger and rumors of unexpected layoffs. I was terrified! How would I be able to pay rent if I were laid off? I began searching for positions within other departments. There were plenty of available positions, but they did not seem to be hiring. I even communicated with various coworkers in several departments who claimed they needed help. The managers were slow at hiring. We were lucky if they even responded to an email. My anxiety level increased, and I became more frustrated with finding job security.

I felt the weight of the world on my shoulders during the next several months. Rent was always due on the first of the month. Gas expenses from an hour commute was burdensome. Casey lost his job at the electronic store. The electricity bill rose each month. The wonderful dream of moving to Charlotte had turned into a nightmare. This was not the life I had expected after college. I became overextended and stressed out, and it wasn't the stress of needing to take a fifteen-minute break and have a cigarette or coffee. I dealt with the stress that kept me awake at night and constantly jittery with intermittent heart palpitations. I visited my Veteran Affairs (VA) physician, and he put me on some blood pressure medication to reduce anxiety. I did not like how the medicine made me feel. I gave the medication the benefit of the doubt and began exercising in the mornings. I also prepared food with low sodium and avoided eating out as much as possible.

Work life became increasingly intense as I witnessed my coworkers quitting or getting fired. My position was stable until my supervisor had to transition into a new department. My new supervisor and I did not see eye to eye. She was strict, impersonal, and humorless. I dreaded going to work at this point. I continued applying to other jobs, but my phone never rang. One day I got tired of being stressed out. It seemed like no matter what I did at work, I was constantly criticized and chastised. My performance began to decline. I did not have the motivation to keep moving forward. New policies and procedures were implemented. These rules complicated our daily tasks even more. Additionally, we had technical difficulties with our computers due to software upgrades for the merger. This environment slowly wore me down.

For the first time in my life, I was depressed with little hope. I developed a bad attitude and desperately scrambled to keep the bills paid and my head above water.

It had been a year since Casey and I moved into the apartment. Our lease was almost up, and I wanted to move to North Charlotte. I found an apartment about ten minutes away from work that I absolutely loved. It was brand new and within proximity to restaurants, entertainment, and retail stores. After moving into my own place, life became simple. It was a scary transition, but it was best for me. I began looking at schools to earn my master's degree. Maybe I could go back to school and make myself more marketable. Meanwhile, I could let the current recession run its course while receiving military educational benefits.

I set up an appointment with an academic advisor at the University of Phoenix. She seemed very impressed with my military background and college experience. She thoroughly explained the Master of Business Administration program to me, and I was immediately sold —not from her explanation though. Just the thought of earning a master's degree was inviting. How could I go wrong with that? It was in my long-term goals to earn an MBA anyway. I just expected to get more work experience before enrolling. We began filling out the application, and I was proud for moving on to a new challenge. Being a student was one area I had mastered since earning my undergrad.

Work life became less engaging as new rules were initiated each week. The level of frustration with software implementations and a diminishing team was overwhelming. I no longer felt appreciated, and my work was taken for granted. I continued going to work because that is what I was programmed to do. After visiting my school to complete some paperwork, I received a phone call from my employment agency. I was told not to go into work that day. I had been fired, but I did not get discouraged. I simply accepted the change in my life. For the first time in 28 years, I was let go from a job. It was confirmation that I needed to move on. With unrelenting amounts of energy, I spent months looking to find another position within the bank with no success. However, I did not feel like a failure. I was actually glad to update my resume with almost a year of solid work experience at a financial institution.

What was I going to do with my time now? I did not feel like a college graduate, and I did not have a job. It was also time to begin making payments on my student loans. I applied for unemployment benefits for the first time in my life. After updating my work experience, I found out that my weekly benefit amount was $50 less than what I was making at my previous job. I was pleasantly shocked with this system and how I was able to remain at home nearly stress-free. I attended one scheduled four-hour class each week. That was the only place I had to be all week. I consistently went to church as well. For months, my Sunday schedule was attending worship service early in the afternoon, going to the mall for lunch, and back home to study or watch television. Mondays through Wednesdays were devoted to reading, studying, and typing papers. Wednesday evenings were dedicated to class, and I was totally exhausted after group discussions, lectures, and sharing ideas. Thursdays, Fridays, and Saturdays were relaxing and eventless. I chose to stay home and save money. I enjoyed this all-new peace and financial freedom. I learned that being a master's degree student required fortitude and additional stress-relieving techniques.

Once my military educational benefits became effective, my money problems faded away. Paying bills became less problematic while I was receiving unemployment benefits and tuition reimbursements. After a year in Charlotte,

I had reached some stability. I could go and eat out at restaurants, buy clothes, and go on road trips. However, I chose to live a frugal lifestyle. I enjoyed financial freedom so much that I wanted to save every dollar. When I wasn't busy with my studies, I played video games, went jogging or simply walked around at the mall. I never expected to experience this type of lifestyle as a student. I did not get too comfortable with it, as I knew it was temporary. I did better financially as a student than I did working full-time, and my bills were paid every month on time.

The Great Recession was the best teacher I ever had. When resources become limited, it places undue stress on people, households, and communities. I thought that I had enough work experience after graduating college. It would have been most beneficial to complete an internship while attending an undergraduate program. However, I could not afford to work for free while attending school. My parents were not in a position to help financially, so I felt accomplished to earn a degree.

Working in retail was not the best experience, but it was necessary. It taught me more about how to deal with the public. Working in hospitality was a step up and taught me how to accommodate traveling business professionals. I often came in contact with people who I perceived were successful. They served as reminders of why I moved to Charlotte. I ultimately wanted to travel and wear suits just like they did.

Moving into the new apartment with my buddy was an awesome experience, and I have fond memories. Life became even better when the position at the bank opened up. I developed hobbies and even felt energetic and alive! The bank merger was a blessing in disguise, and I should have never been so stressed out. God had a plan and tested my faith. After moving into my second apartment and beginning post-secondary school, life became simpler. I learned that being fired from a job didn't mean I was blacklisted for the rest of my career. It just meant I had to change my approach and perspective on life. There are always opportunities around you. You just have to open your eyes to positions and people. There is power in networking with people on a regular basis.

Developing a weekly routine helped me to set a pattern. Going to class and church gave me constant contact with people and positive energy. When you are trying to accomplish a goal, it is important not to isolate yourself. Although you may need to concentrate most of your time alone to work or study, you must spend some time communicating in person with others to release toxic energy.

Those six years of military experience paid off when it was time to attend undergrad and graduate school. The military educational benefits helped propel me to an unprecedented level. I was able to focus my time concentrating mostly on completing school. When I was 18 years old and became active duty, $100 was deducted from my paycheck each month for a year. It was a worthy investment that took years to come into fruition. That was evidence that planting seeds early yields sustaining life benefits in the years to come.

Chapter 7- A Different Perspective

After my second military educational benefits disbursement, I planned a three-day trip to Paris, France. I had always wanted to travel internationally but did not have the time or money. Those resources eventually became abundant. I flew to Paris for New Year's Eve in 2010. That was my first time leaving the United States after separating from the military. When the plane touched down in Paris, I was astonished by hearing French everywhere I went. Maybe I could now put those French classes I took in high school and college to work. I took the train from the airport to Gare de l'Est station. I grabbed a taxi and found my hotel in this area. Although I was exhausted from the six-hour flight, I was eager to get out and enjoy the city.

I found it easy to communicate with the French people. I was expecting them to be blunt, rude, and disrespectful. However, they were the total opposite. Parisians were patient, welcoming, and respectful. I did not understand. All my life in my home country, I was taught to believe French people were snooty. How did this happen? I felt bad for labeling these lovely people, and they treated me better than most Americans. Maybe the media in the United States attempts to control how we view other countries and the world. Maybe French people are just nicer now. I am not sure what the explanation was, but I was blown away by their generosity.

Checking into my hotel was a breeze. The room was a bit smaller than I anticipated, but I did not plan on spending a lot of time there anyway. I headed out in town to eat. I was looking forward to eating an authentic croissant! To say there were a lot of pastry shops around would be an understatement. Seriously, they were everywhere. How did the French eat so many sweets and stay slim? I was able to answer this question later in the day. They walk around the city a lot. In the United States, we drive cars most of the time, unless you are in a huge metropolis. In Paris, they walk, ride the bus, ride the train, and take taxis. I took pleasure in scouting the city and exploring. It was cold outside, and I was prepared as I brought wool sweaters and a peacoat. Many Americans would not have appreciated visiting Paris during this time of year. However, I embraced it and felt on top of the world. Thoughts began swimming through my head about finding employment in France after graduate school. I wondered what it would be like if I met a Parisian network of acquaintances and colleagues.

I visited La Basilique du Sacré Coeur de Montmartre and was amazed by its beauty. The church sat on top of a hill providing a magnificent view of the city. It was a huge attraction, and people were everywhere. There were tourists, musicians, and merchants who made this a memorable experience. After walking the city for hours, I spotted the tip of the Eiffel Tower. I began walking toward it thinking I could be there in about 20-30 minutes. It seemed like I was walking forever, and it even got slightly bigger; then dusk began to set in. I ended up hailing a taxi to take me to the Eiffel Tower before it got too late. I stood beside the iconic French monument in total awe. It was tall, well lit, beautiful, and breathtaking. It was much bigger in person than I ever imagined. I paid for a ticket and proceeded up the elevator where I saw a brilliant skyline. It was cold and windy at the top as tourists chatted and pointed fingers toward the ground. I heard someone speaking English and made it my business to introduce myself to her. She was an American teacher working in Paris! So, it was not just a dream. Americans traveled internationally all the time. She explained how she was enjoying her teaching experience in Paris. Her words and enthusiasm motivated me. I felt encouraged to keep working hard so I could travel internationally and get paid to do something I love.

The next day, I spent more time exploring the city. I visited shops and restaurants. The best part was interacting with the locals. Bartenders usually spoke a little English, and I could hold a general conversation with them. The wine was always flavorful, and the cuisine was astounding. I had no complaints about Paris. I enjoyed some nightlife later on. I was not sure where nightclubs were, so I jumped into a taxi and asked the cab driver to take me to a place. I finally arrived in a social atmosphere where I could speak freely with others and not feel awkward initiating conversation. I met people from abroad, and they were laid-back. We took photos, sipped on some drinks, danced, and laughed. I remembered experiencing life's lowest moments after graduating college. Now, I was in Paris partying with friendly people and enjoying life to the fullest. How did the paradigm shift? The previous year I had to wonder how I was going to eat lunch. Now, I was halfway around the world surrounded with gourmet food and fine wines. After getting some contact information to connect with my new acquaintances online, I left and spotted a Ferrari parked right outside of the club. As a car enthusiast, I was truly living a memorable experience.

When attending jazz events back home in Charlotte, I met an older gentleman, and we had a lot in common. He was a traveler too. He had a contact in Paris and encouraged me to connect with her. I sent her a message on social media, and she responded. We met up for a friendly dinner at a small restaurant. I liked listening to her cute French accent. Now, I had new acquaintances in France who could potentially be long-term friends and resources for useful employment or housing information.

On New Year's Eve, I spent the evening eating and drinking wine at a restaurant with another American. He was a lawyer from Florida and was visiting his fiancée. We had a blast until midnight, and then I kept hearing "Bonne Année," which is Happy New Year. I heard it in the restaurant, on the streets, and in every establishment. Champs-Élysées, a famous street with shops and boutiques, was swarmed with locals and tourists celebrating. The Eiffel Tower glittered with its sparkly lights. It was an awesome and spectacular night!

The next day was quiet in Paris. Many restaurants and places of business were closed on New Year's Day. I spent the day taking photos and savoring more food and wines. My flight home was early the next morning. I did not sleep well that night in fear of missing my 7 a.m. flight. I set the alarm on my phone and sort of cat napped though the night. The next morning I was so tired and hoped to snooze on the plane. All of the walking and late nights had caught up with me. I quickly got my belongings together and checked out of the hotel. Time began to slip away as I scurried to the train station. After arriving at the airport by 6 a.m., the lines were ridiculously long. There was no way I was going to catch my flight. Three hours and $350 later, I found myself on another flight back home. I just wanted to land on U.S. soil, drive home, get settled in my apartment, and lie in bed.

Arriving home felt like a relief after dealing with the busy airports. I was content relaxing in my apartment where I was comfortable. I began looking for overseas opportunities online during my spare time. It would be wise to work in an English speaking country though. I had about a year until I was finished with my MBA and needed to make an informed decision.

I spent the entire year of 2011 focusing on my studies. I marveled at the ideal of working and living abroad. Periodically, I would watch videos and look at photos of exotic places online during my free time. I remained on track by motivating myself about the possibilities after school. I was open to going almost anywhere in the world to find my calling in life. I began to prepare and determine where I would migrate after graduation. I spent the next six months analyzing various countries, their cultures, and job markets.

Visiting France gave me a different perspective on life. This trip taught me to not only think about my city or the U.S. but to think on a global scale. The importance of learning a second language was huge. I did not know French fluently, but I knew enough to get by in Paris. If I had not known conversational French, the locals probably would not have received my presence well. However, a few French words and phrases welcomed red carpet treatment in France.

Although they like pastries and chocolate, they keep in shape by incorporating walks and exercise into their daily routine. In my opinion, healthy people live longer, are happier, and seem to make more money. I noticed they had a strong sense of pride in their culture and presented themselves well in public. In the U.S., it is common to find an American at the grocery store in pajama pants on an early Sunday morning. The French people were very stylish and, by far, the proudest people I have had the pleasure of visiting.

Seeing La Basilique du Sacré Coeur de Montmartre, the Eiffel Tower, and Champs-Élysées gave me a chance to visualize life from a French perspective. They have beautiful scenery throughout their country and appreciate it just as much as Americans do the Statue of Liberty. Whenever I walked into a shop or restaurant, I was greeted promptly and given superior customer service. They have specialty shops unlike most corporate retailers like the U.S. I walked into a pharmacy to buy hand lotion, and the attendant answered every question imaginable about the moisturizing creams she had available. When going to restaurants, the servers and bartenders personalized my experience by being friendly and efficient. They made sure my glass did not go empty and provided a check promptly when I was ready to leave. These were the small, kind gestures that make a strong, thriving economy.

I was surprised to see so many Americans abroad in Paris, but they were all over. I usually did not recognize them until I heard them speak. Once again, I had hope that I could deviate from the Americanized way of thinking and discover a new approach. Partying with the French was a life-changing experience. Although they like drinking wine, they are very friendly when there is a party. They are creative and festive people who enjoy a great time. Traveling to Paris allowed me to see the world from French lenses, and I truly understand how their culture has made a positive, worldwide impact for centuries.

Chapter 8- A Down Economy

It was 2011, and the U.S. economy still had not completely recovered. After three grueling years, good-paying jobs were scarce unless you had extensive work experience. It seemed like the only available jobs were low-income, entry level or high-income, and experience-based. The entry level jobs allowed a person to gain basic knowledge of an industry and provided a "foot in the door." However, the pay was so low that you had to consider finding a roommate to balance out living and transportation expenses. Since it had been years since the economic collapse, I began to explore other career opportunities. Despite all that occurred, what could I do to become a success? How could I combine my military, college, and work experience to increase my job marketability? The military taught me that I had to relocate to complete assignments. College taught me that I must have a system in place to generate income. This system must be replicable, adaptable, and accessible. Plenty of work experience taught me how to relate to a multitude of people—young, old, poor, rich, country, and city folks. Since the job market was slowly emerging, I considered the possibility of being an expatriate. There must be a company somewhere in the world that needed my brains and will power. After contemplating many countries, Australia was attractive because of its similar values to the U.S. The best part was that they spoke English, and I would spend less time adjusting to their culture. I used my birthday as a deadline: I told myself, "If I have not found a conditional job offer by July 1, I will relocate to Australia to build a new life." My mindset was focused on doing something over the top, out of this world, and completely out of the box.

I continued searching for jobs in the U.S. for six months but to no avail. After much prayer, that was my God-given sign to move forward with Australia. I began conducting research online about Australian history and culture. I looked at videos, online images, and used social media for viable information.

After a month of careful consideration, I purchased a work holiday visa in Australia for a few hundred dollars. I was officially invested in relocating to another country. A couple of months later after more extensive research, I purchased a one-way plane ticket to Sydney for $1,300. My intentions for international work experience became plans and were no longer a forethought.

I spent the remainder of 2011 focusing diligently on my studies. My routine stayed the same. Graduate studies were so intense that I looked forward to the weekends to decompress. Earning a master's degree is expensive and requires a tremendous amount of dedication over time. However, it can be done with a plan and discipline. There were plenty of days that I wanted to set my laptop to the side and go do something fun. As the year came to a close, I began making arrangements for ending my apartment lease, selling my household items, and selling my car. Meanwhile, I continued searching for jobs in Sydney so I could have a secure position before my arrival. I learned that Australian employers required your presence when applying for jobs. Without an Australian address or phone number, it was practically impossible.

My last class of earning my MBA was in-depth, fun, and exhilarating. I developed a business plan and presented it by PowerPoint in class. There was little group activity at this stage. My professors gave me the basic tools to use my advanced degree in the real world. These tools taught me how to be a producer. My analytical skills had drastically improved, and I learned how to monitor, maneuver, manipulate, manage, and maintain a business! After successful completion of all my classes, I earned a 3.7 GPA.

Moving to Sydney was nerve-racking, but I expected it to thrust my career forward. I wanted my family to witness my bold move to achieve an unimaginable level of success. I felt the pressure to pave the way for them. I was an MBA graduate, and they had great expectations. I had the power to change their lives as well. After selling most of my household items and my car, I boarded a plane with a huge duffel bag and flew 10,000 miles around the world for an epic adventure. It was time for the biggest exploration of my life thus far.

An underperforming economy will do one of two things to people in the work force. It will either make them work harder or give up. Inflation, increased prices of gas, groceries, and energy make it even tougher. Entry level jobs that pay $10-12 an hour usually do not pay all the bills. I had to use what I had to move forward. The three areas that made me unique were my military, college, and work experience. Additionally, I had traveled and could relate to foreigners. These attributes added value to my job marketability. I just had to find a creative way to convey that high level of marketability to employers. I had an extraordinary mix of accomplishments I could use to sell myself.

Your presentation during an interview is what sells employers, not so much the way you look. You must look professional, but hiring managers are more interested in how much positive energy you can bring to the team and the company.

Setting deadlines helped me to complete my MBA program and prepare for my international relocation. As I prepared for my voyage to Australia, I gathered as much data as possible. I looked at maps, studied the areas, and gained an idea of what to expect before I even stepped foot off American soil. Taking reasonable leaps of faith gave me the courage to uproot my entire life and start over in an unfamiliar place. For most people, that would be petrifying. However, I did not have much to lose during a down economy, and I had always wanted to be an expatriate.

Chapter 9- Expatriate Life

The flight to Australia was 23 hours long. It started in Charlotte, and I had a layover in Los Angeles. Life seemed to be in my favor and moving in the right direction. I walked through the airport with high self-esteem and determination, as I mentally prepared myself for relocation Down Under. I was leaving the country and pinched myself to see if I was dreaming. I boarded my international flight with no hesitation and felt free once the tires lifted off the ground. I occupied my time flying across the Pacific Ocean by sleeping, reading, listening to music, and watching movies. The twelve hours went by relatively quickly. After landing in Auckland, I awaited a 90-minute flight to Sydney. When approaching the airport, I looked out the window and saw the Opera House and Sydney Harbor Bridge. This place was real, and I was finally there! It was fascinating to leave North Carolina in the winter and arrive in Sydney during the summer all within a day. After grabbing my bags, I proceeded to the train where I traveled to

Circular Quay station. I jumped in a taxi and went to check into my hostel in The Rocks, an historic area in the city. After getting settled into my room, I hailed another taxi and went to the mall so I could purchase a phone. This was awesome. I had an official Australian address and phone number. Now I could begin applying to jobs. I spent the remainder of my first day exploring the city and taking photos. I was overwhelmed with excitement just walking around and taking in the new environment. The locals were refreshing to speak with, and I was considered the foreigner with an accent. Clouds began to set in when I finally made it to the Opera House and Harbor Bridge. My dream had come true, and I was standing in one of the most remarkable and iconic places in the world. It began raining, but I continued snapping photos and smiling.

Waking up the next morning in Sydney was nothing less than spectacular. I was now rested and ready to explore more. I had all the time in the world to walk around and enjoy the sunny, warm weather. I certainly was not used to summer in February. The restaurants and cuisine galore was enough to draw anyone into the Aussie culture. Although I enjoyed my first few days getting settled in, I eagerly wanted to see what the job market was like. I was willing and ready to submerge myself into this new, mysterious Land Down Under. I started by scouring online through employment agencies for a position. I wanted to protect my future and travel investment by getting a job to defray costs. I began submitting my resume online with several offices. While waiting for a response, I did recreational activities such as jogging across the Sydney Harbor Bridge, going to Bondi Beach, and trying so many amazing restaurants. I felt free. I felt so liberated. I felt like all the hard work thus far in life was worth it. My six years in the Navy, bachelor's degree, and MBA had paid off. I was on the most beautiful island in the world with very hospitable people. I was impressed with the Aussies. They were very friendly, good-spirited, and warm-hearted people.

It had been a week, and I began to get anxious about finding a job. Sydney is also not the cheapest place to live. I knew that the sooner I found a job, the better off I would be. I also did not want to deplete my savings. I began meeting people and getting acquainted with locals. I wanted to enlarge my circle of friends, and networking is always a fun way to accomplish this goal. I finally received a phone call for an interview at a marketing firm. This was my chance to shine and make a difference!

I dressed in my best threads and showed up to my interview. There were many people and an energetic aura within the office. I filled out some paperwork and prepared for a group interview. I sat there with new faces and felt engaged. We discussed past sales experiences and what made us successful. The interviewer was impressed with my feedback and requested that I stay for a second interview. I was grateful and waited patiently. However, he had not exactly explained the job description. I was just thrilled with having a job opportunity. After finishing the second interview, the hiring manager asked me to go out with a salesperson into the field to observe. I still was not completely clear on what he would be selling. After hanging out with my new acquaintances, I quickly gathered this was a door-to-door sales position. I felt reluctant, but these guys were so cool and easy to get along with. We simply were attempting to switch consumers to a different energy provider. I was a walking and talking commercial. I was a little embarrassed performing these job duties since I had just graduated with a master's degree. However, I remained humble and remembered this was a starting point. So, I accepted the job offer and was considered a contractor for the marketing company.

Every morning, we met at the office and received training. They played upbeat music and made us feel appreciated. We were placed in small groups and sent to different parts of the city. I was not thrilled with knocking on doors for a living, but if I did well, I could make $1,000 a week.

We did a lot of walking—so much walking that my legs and feet hurt. It was hot, and we had to wear shirts and ties as well. I remained optimistic, and it felt great going to work each day. Maybe I would meet other people who could get me involved with an employer who had a desk position available. I knew that sitting around doing nothing would produce zero results. So, I embraced my position and coworkers with the hopes of achieving incremental goals.

For several weeks, I got up each weekday and commuted to work. While living on Bondi Beach, I took the bus and train to work. While living in Pyrmont, I rode the trolley, which only took fifteen minutes. Finally, there was a work routine and daily schedule. After our daily training, my coworkers and I would sit at the coffee shop on George Street near Central Station and have breakfast. I usually ordered an egg and cheese croissant and orange juice. After work, we hung out at the bar close to the office and shared a pitcher of beer. I truly felt a part of the team. There had been very few times in life when I felt this accepted. The Australians did not care what kind of car I drove, where I lived, or how much money I had. They did not even mind that I was an African American. I was eager to build a life there, and I was going to make it happen one way or another.

After work and on weekends, I entertained myself. I went to restaurants, nightclubs, and out with friends. I even went out on dates to understand Aussie women and Sydney culture. I met a Brazilian woman on the bus heading to work one day. She had the most beautiful brown eyes and carried great conversation. She and I were 30 years old, and she was an architect. She had a jazzy personality, and I could listen to her talk for hours. One night, we went to a food festival and had a blast. On a separate occasion, we had a romantic date with dinner, drinks, and a walk. This was one of those nights when I really connected with someone and felt the magic. The weather was fantastic, and we shared many laughs and even held hands. She was my world at that moment and gave me butterflies in my stomach. This was how dating should be—fun and adventurous. I did not feel any pressure to impress her with money. It happened naturally, and our time together faded away quickly.

I usually woke up each workday with my game face on. I think that is part of being a sales professional. I had to pump myself up and ignite a fire inside myself to be successful. It was March 16, 2012, and I woke up early so I could call home. I got dressed and commuted to work. I had been with the marketing firm for about a month and felt the need for improvement. Not only did I want to make a lot of money, but I wanted to impress my boss and fellow coworkers. The pressure was on, and my sales were down for the week. A true sales professional understands that his job is simply a game of numbers.

It was my plan to cover a lot of ground and houses this day. My team and I arrived in a neighborhood and split up. As I walked house to house, nobody seemed to be home. This was a problem because I was determined to meet my goal. I began to get impatient because it was a hot, exhausting week, and I was expected to work a Saturday since my numbers were down. I finally approached a home where the door was opened. After about 20 or so houses, there was a glimmer of hope. However, there was a gate, and it was locked. This was a point in my life when I should have just kept walking, but the fire ignited within was far too great to walk away. I set my backpack and iPad down on the concrete gate, hopped over it, and walked up to the door. This was one of my bravest moments, and I would have never done this at home in the United States where many homeowners have guns. I knocked, and a man came to the door. He immediately asked how I got through the gate, and I gave a witty response to distract him. He did not crack a smile and did not seem impressed. He was not eager to talk to me and politely asked if I wanted him to open the gate so I could leave. Once again, Australians were super friendly even after trespassing onto their property. I should have accepted his offer to open the gate. Instead, I was bold and attempted to rescale the gate. As I approached the gate, I began running to propel myself over the gate. I did just fine until an inch of my shoe caught on the concrete in midair. So, there I went five feet in the wind when I felt my body losing balance and falling forward. Everything seemed to move in slow motion, and I could see my face impacting the ground first. My only reaction was to put my arm around my face. Upon impact, my first thought was, "This isn't bad. I can just get up and dust myself off." I saw my elbow bounce off the ground and into the air at a weird angle. My arm felt a bit tight, and I stood up. Then there was extreme pain and agony. It was excruciating, and I had never felt pain like this before. I could not help but yell when a neighbor and coworker heard me. They came to my rescue and walked me into the neighbor's home. She was very hospitable and offered me a glass of water. They called an ambulance and hauled me off to the hospital where I had to undergo surgery. I was upset because I had plans that night. Medical personnel made it clear that I was not leaving that night. The next day, I woke up with a cast on my arm. I felt a lot better, but my arm was sore. I had never broken any bones as a child or in my adulthood. The doctors discovered that I had a complex elbow dislocation. I also fractured my forearm and tore a ligament. I had flown 10,000 miles and gotten hurt, and it was my fault for jumping that gate.

My parents were panicked when they learned about the incident. I kept everyone informed on social media about my condition and called home to speak to my mother who acted like I had a near-death experience. I was only concerned with healing quickly so I could get back to work and making money.

The whole purpose of traveling to Sydney was to use my MBA to find a job so I could start a new life. This injury was a major setback—physically, emotionally, and spiritually. My body was tired. I was angry with myself for not making better decisions, and my spiritual health began to suffer. I returned to work a few days later. The job was physically demanding. Now I had a hot, sweaty, and itchy cast on, which forced me to work with one arm. It was tough even to sleep, shower, and eat. My friends encouraged me, but my dream was broken; soon I began to run out of money. The next day, I chose to quit the marketing job. It really took a toll having to let my boss and coworkers down.

I began to get homesick, which really frustrated me. I was in paradise and had wanted this experience for nearly a decade. I no longer felt secure, and it was tough working with a handicap. I found another job at a Mexican restaurant as a bartending assistant. It was not the most generous salary I was expecting after graduate school, but it was a modest paycheck. Once again the people I worked with were friendly and outgoing. The job was only part-time though, and I did not know how long I could sustain this type of work. During my off time, I made numerous phone calls, sent dozens of emails, and followed up with recent connections. There were jobs available, but employers were looking for applicants with Australian citizenship. My professional experience Down

Under was slipping through the cracks. I honestly did not know what else to do. I never stopped looking for a white-collar position.

My arm began healing, and I felt a lot better. I looked forward to nighttime and sleeping. Nights were quiet and peaceful. When the sun rose, it was time to hustle and bustle or meet and greet with people. Due to limited work hours at my bartending job and my inability to find full-time employment, I could no longer afford to stay at the hostel. I went to the U.S. Embassy and spoke with an ambassador about returning home. It was a humbling experience, as I had to request federal financial assistance for airfare. I stayed with a friend my last few days in Sydney. It was a relief to arrive at the airport. However, I felt defeated. I felt overwhelmed and like I had lost a huge battle, but this struggle was over. I could go home and start over again.

My overall expatriate life in Sydney, Australia was extraordinary. I had an awesome opportunity to travel to another country to live and work as I had desired for years. I visited Australia for the first time in 2001. I was active duty and on deployment when our ship's captain announced we were visiting Perth and Hobart. I never forgot my experience there and always wanted to return. I felt fulfilled knowing that I had lived my dream. Experiencing this newness was necessary. It gave me a chance to get away and see how others live life. Although the U.S. and Australia share similar cultures, I noticed a difference in how the countries conduct business. Australians carried out business practices in an indirect, laid-back manner. American business practices are more formalized, direct, and aggressive in nature. I found myself a little frustrated on occasions when trying to get official business done in Sydney.

Living the city and party life in Sydney was epic. There were always social events and festivals going on somewhere. Fun, dynamic, and enthusiastic people were contagious. The transfer of positive energy was amazing, and that is the one aspect of Australia that I will always miss. Gourmet food was easily accessible. Anywhere I walked in the city, especially Darling Harbour, I smelled exquisite food. Since we were on an island, seafood was fresh and perfectly seasoned. I remember hanging out with my friends and looking into the star-glimmering night sky thinking I was in one of the best cities in the world.

After being in a foreign country for two months, it was tough working a commissioned sales job and finding a private dwelling. I found refuge in hostels most of my time there, although I wanted an apartment. After dislocating my arm, my priorities changed since I was not making the desired salary there anyway. The trip was an experiment, and I did not have much to lose. I could always go home to the U.S. and find a job easier. I appreciated my trip Down Under and would not trade the experience for anything in the world. I felt at ease flying back to the U.S. and knowing that I could scratch an item off my bucket list.

Chapter 10- Defining Your Success

It felt incredible arriving back in the U.S. It was empowering to land in Los Angeles where American life was familiar. I had a connecting flight to Charlotte and was looking forward to seeing my family. After getting settled in, my first agenda was finding a job. I chose the shotgun approach and applied to as many jobs as possible. I graciously wanted the phone to ring the next day, but it did not. My phone did not ring the next week either. It was a month before I heard from a potential employer. I went to an interview for a position at a furniture store. It was a sales/collections position, and I accepted the job offer. I was not thrilled about working in sales again, but other potential employers neglected to offer me a position. It was an adjustment calling delinquent customers and asking for payments. However, God was planting a seed for me. At the time, it was just a job, but the collection and negotiation experience became very valuable.

I worked in that position for seven months before transferring to a different store. My sales skills had improved remarkably. My collection soft skills had improved as well. My performance exceeded my manager's expectations. I closed sales just as well as she did, and our district manager was impressed. We worked as a team, and our store frequently ranked above average within our district. Maybe I had found my niche. Maybe my hard work would be recognized and lead me to a higher position. Maybe I could move to Dallas, Texas and work for the corporate office. Endless possibilities went through my head. As I approached a year with the company, a store manager position opened within the area. It was a new store, and I knew with certainty it was a chance for me to move up. After expressing interest with my district manager, he encouraged me to apply online. If I were offered this position, it would boost my confidence and self-esteem. The professional experience would also look great on my resume. I deserved a pay raise and put my faith into a new beginning. After the job posting closed, my district manager's delayed response was unnerving. Maybe he was just thinking it over. I refused to let any negative thoughts creep into my mind and, instead, was determined to get this position. He visited my store with an offer. However, it was not the offer I was expecting. To my surprise, he offered me a position in recovery as a repo man. I was shocked and disappointed. He claimed that was the best position for me based on my personality and character. However, I knew my self-worth. I had no experience driving a truck and picking up furniture for delinquent accounts. Although I had a year's experience sitting by my manager's side, it did not seem to be enough. My education and degrees did not seem to matter. My district manager's decision was based on who he thought I was or what he wanted me to be. I was sure office politics had played a role in his decision. The turnover rate for the recovery position was high, and I felt more secure in the office.

After our meeting, I felt disheartened. I thought, "Why can't they think twice about how much of a loyal employee I am?" I applied for a sales position at a car dealership shortly afterwards. I did not want to continue working for a company that gave me false hope. I left the furniture store and focused my hard work and attention elsewhere.

The car dealership welcomed me with open arms. I always loved working around cars. I learned the vehicles, models, and specifications quickly. Since I had mastered sales in the furniture store, it was time to up the ante. Working in sales simply means understanding people. When you understand their problem, you can offer a solution. However, markets are ever changing, and consumer behavior is different based on your location, industry, and consumer expectations. I spent more time than most other sales people getting to know my customers and connecting with them. I knew that if they felt that I truly wanted to solve their problem, they would be more likely to do business with me. Our economy was still slow, which directly affected sales. Many people could not get financed due to poor credit, unstable income, or wanting more car than they could afford. Most of all, customers were looking for the best deal, even if they did not like the competitor's car. It was a 100% commission job, and my pay began to suffer after three months. After my apartment lease was up, I chose to move back home with my mother in South Carolina. This was my last resort. However, I had to swallow my pride and find a more stable, decent paying job.

I continued my search for a position that paid a higher wage and aligned with my experience. Going back to the furniture store was not an option. I picked up a part-time retail job as I applied to more full-time positions. I thought about starting my own business, but what would I sell? The only service I had provided in the past was auto detailing. It satisfied me to see a dirty vehicle and transform it into a shining beauty. It was even more satisfying to see a customer's pleasant reaction. The downfall to auto detailing was its seasonal nature; many customers consider it a luxury, not a necessity. As long as it was warm outside, customers would purchase this service if they had disposable income. But what would happen in the middle of winter when demand was low? I needed to invest my time in a year-round service if I wanted to survive as an entrepreneur.

Fortunately, I kept in touch with a buddy and former coworker from the furniture store. He was working at a popular, high end grocery store as Assistant Inventory Manager and encouraged me to apply. I was not thrilled with working in retail again, but it was the best pay I could find at the time, and there were plenty hours available. I interviewed and accepted the job offer. I have always done exceptionally well with interviews. My secret was having a meeting of the minds with my interviewer and exuding a high level of confidence that I could do more than just a good job.

I worked nights at the grocery store and days at the toy store. I did not sleep much. However, I was able to move back to Charlotte and buy a cheap $500 car for transportation. I shared an affordable three-bedroom condo with two other guys in a prominent area. So, life was picking back up. I continued working days and nights for three months. I may have slept two or three hours between jobs, and I felt miserable some days. However, it felt so good to be free and on my own. Having patience and fortitude was the key. I had not lived a traditional lifestyle of graduating high school, going to college, starting a job, and getting married by the time I was thirty years old. I placed a lot of unnecessary pressure on myself trying to live up to these expectations all these years. Although I rented a small room and drove an ugly, old faded wagon, I was content. **My value system had shifted, and my survival was my success!**

After Christmas of 2013, I discontinued my seasonal, part-time job at the toy store. Another full-time opportunity became available three months later, and I submitted my two weeks notice at the grocery store. I interviewed for a collections position and accepted a job offer at an auto finance company. I finally had my own desk and computer again.

Why was this position not available when I initially moved to Charlotte in 2009? Maybe it's because it was not a part of God's plan for me. Adverse conditions challenged me, and made me stronger; if I had never been stretched, I never would have grown. I eagerly trained for my new job, appreciated my workplace, and enjoyed the presence of my coworkers. My preconceived image of graduating college and starting a job with $40,000 salary was a bit far-fetched during an economic recession. My expectations became even more unrealistic and distorted after earning my MBA and looking for a $60,000 salary job. It is possible to achieve that level of income as a new graduate, but the opportunity did not present itself; I am fine with that now. When you train your mind to believe more income is achieved with more experience, it will free you from expecting too much too soon.

The best aspect of my job is the relationships I have with coworkers. My advice for new graduates and post-graduates is once you find a workplace and position where you belong, be patient and endure in it. Advancement will come in time as you network and sharpen your knowledge, skills, and abilities. So, was college a curse? No, it was not. However, my extreme expectations afterward college were. College taught me life skills, analytical skills, and teamwork. It felt like a curse because I tried to muscle my way into Corporate America. I aspired and craved the prestige of wearing a business suit, carrying a briefcase, and having a corner office. If I could return ten years in time, I would have told the uneducated version of myself to relax. It is great having a general plan, but attempting to map out every part of your life will stress you out. Enjoy where you are in life right now because it will evolve, and you will remember the good times when your situation changes.

Teaching yourself to be flexible is a wonderful quality, and it allows you to adapt to any environment. When reminiscing about my military, college, graduate school, travel, and work experience, I was already a success and did not even know it. I was so caught up in the materialistic American Dream that I lost focus of what really matters like family, good health, and making memories with those you love. Ten years from now, people are not going to care what clothes you wear and what kind of car you drive. They will care more about how you positively influenced their life and how well you made them feel. Build relationships everywhere you go, whether it is down the street at the grocery store, across the country, or around the world. That positive energy will always come back and find you. Learn to love, endure, embrace change, build relationships, and success will find you despite where you are in life.

Defining your success is a challenging task. Everyone defines success differently. Some people define success as having lots of money. Others define it as having lots of time with family. We may even define success as owning businesses and real estate. Our definition of success changes over time as we experience life. While learning to define my success, I learned it is ok to start over. Let's face it—we make mistakes and misjudgments. However, that is not an excuse to give up because too many people have invested in your future.

Take opportunities as they become available. They lead to bigger and better opportunities. You may explore different interests and industries, but remember longevity in the workplace evokes stability. We often take jobs to fulfill financial obligations until we can work in our desired field. During this process follow your heart, not Corporate America. Having a strong sense of self-awareness will change your life. Observe what goes on around you at home, work, and school. Surround yourself with positive, uplifting people who will hold you accountable when possible.

In the workplace, understand your customers and build relationships. Going to work can be stressful like you are going to war every day. Relieve yourself of daily stress by exercising and eating a healthy diet. Your body will be far more energetic and productive when you eat a balanced meal and get plenty of rest. As you go about your day, remember your survival is your success. I discovered that people, numbers, and cars are my niche. It feels wonderful to walk into a job that I enjoy every day. Use your gift and heart's deepest desires to spread positive energy. Albert Einstein once wrote, "Energy cannot be created or destroyed; it can only be changed from one form to another." So when you are faced with adversity, think positively and find a way to turn your lemons into lemonade. Most importantly, trust God. Believe in faith, not sight.

Made in the USA
Coppell, TX
23 May 2021